spot

EXTREME WEATHER

TORNADOES

by Anastasia Suen

AMICUS | AMICUS INK

shelter

funnel cloud

Look for these words and pictures as you read.

supercell

poles

Uh-oh. A big storm rolls in.
It could make a tornado.

See the supercell storm?
It pulls warm air up.
Cold air falls down.
This makes the clouds spin.

funnel cloud

See the funnel cloud?

It spins fast.

It is made of water drops.

The cloud touches the ground.
Now it is a tornado.

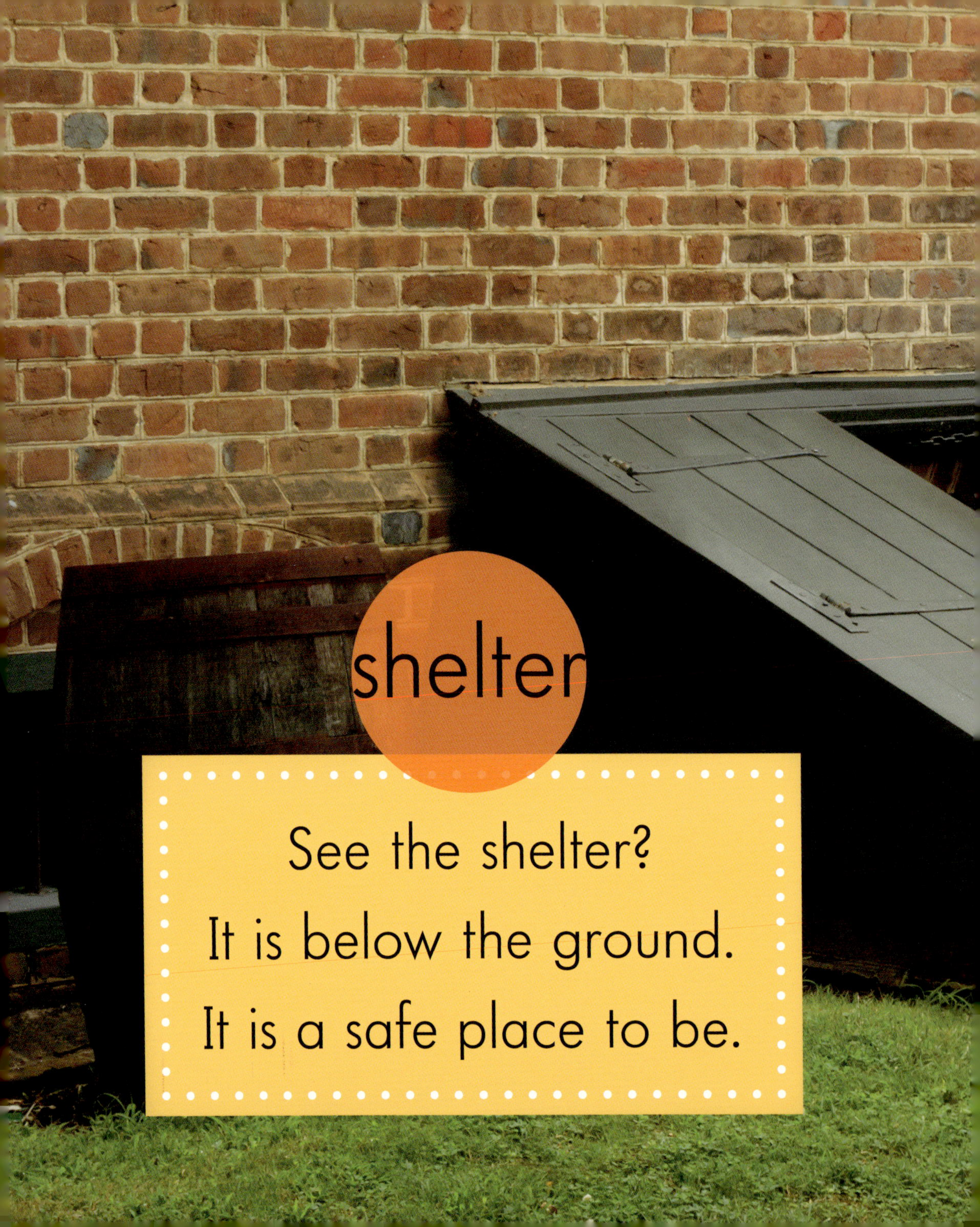

shelter

See the shelter?
It is below the ground.
It is a safe place to be.

poles

See the poles?
The wind is strong.
It pushes the poles over.

The roof came off.
But the family is safe.

shelter

funnel cloud

Did you find?

supercell

poles

Spot is published by Amicus and Amicus Ink
P.O. Box 1329, Mankato, MN 56002
www.amicuspublishing.us

Library of Congress Cataloging-in-Publication Data
Names: Suen, Anastasia, author.
Title: Tornadoes / Anastasia Suen.
Description: Mankato, MN : Amicus, [2021] | Series: Spot. Extreme weather | Audience: Ages 4-7. | Audience: Grades K-1.
Identifiers: LCCN 2019037110 (print) | LCCN 2019037111 (ebook) | ISBN 9781681519456 (library binding) | ISBN 9781681525921 (paperback) | ISBN 9781645490302 (pdf)
Subjects: LCSH: Tornadoes—Juvenile literature.
Classification: LCC QC955.2 .S84 2021 (print) | LCC QC955.2 (ebook) | DDC 551.55/3--dc23
LC record available at https://lccn.loc.gov/2019037110
LC ebook record available at https://lccn.loc.gov/2019037111

Printed in the United States of America

HC 10 9 8 7 6 5 4 3 2 1
PB 10 9 8 7 6 5 4 3 2 1

Alissa Thielges, editor
Deb Miner, series designer
Ciara Beitlich, book designer
Marty Levick, photo researcher

Photos by Getty/www.dennisoswald.de cover, 16; Shutterstock/Minerva Studio 1, 8–9; iStock/PeopleImages 3; iStock/mdesigner125 4–5; Alamy/Design Pics Inc 6–7; Dreamstime/Travis Rogers 10–11; Shutterstock/Everything I Do 12–13; iStock/gmcoop 14–15